MARK HOOPER

Buy That Classic Car You Always Wanted

Why to Buy, What to Buy, How to Buy and the Life-Long Joys of Ownership

Copyright © 2024 by Mark Hooper

All rights reserved. No part of this publication may be reproduced, stored or transmitted in any form or by any means, electronic, mechanical, photocopying, recording, scanning, or otherwise without written permission from the publisher. It is illegal to copy this book, post it to a website, or distribute it by any other means without permission.

Mark Hooper asserts the moral right to be identified as the author of this work.

Mark Hooper has no responsibility for the persistence or accuracy of URLs for external or third-party Internet Websites referred to in this publication and does not guarantee that any content on such Websites is, or will remain, accurate or appropriate.

Designations used by companies to distinguish their products are often claimed as trademarks. All brand names and product names used in this book and on its cover are trade names, service marks, trademarks and registered trademarks of their respective owners. The publishers and the book are not associated with any product or vendor mentioned in this book. None of the companies referenced within the book have endorsed the book.

First edition

*This book was professionally typeset on Reedsy.
Find out more at reedsy.com*

"Luck is what happens when preparation meets opportunity"

ROMAN PHILOSOPHER SENECA

Contents

1	Introduction	1
2	Why Buy a Classic Car	4
3	What to Buy	7
4	How to Buy	13
5	The Joys of Ownership	18
6	Conclusion	22

1

Introduction

Welcome to Buy That Classic Car You Always Wanted!

My name is Mark Hooper and I'm excited to share with you the reasons why I bought classic cars, what I have purchased and how I went about it. I cannot express to you how exciting this journey is and how fulfilling it is to take this path and make your dream come true.

I have loved classic cars all my life. When I was a child I was drawn to long, low, elegant and powerful American cars, which were few and far between living and growing up in Australia. My lust for beautiful cars stemmed from books, TV shows and movies, plus a handful of Australian built luxury cars on our roads in the 60s and 70s. Remember there was no Internet back in those days, so my resources and exposure was way more limited than today.

No doubt you are here for the same reason as what inspires me – you just love classic cars! To me, they are all exquisite in their own unique way and are quite literally works of art on wheels. From the simple engineering, to the iconic and inspiring lines, topped with lashings of chrome, who can argue they make your head turn more than ever today. Sure, you have a daily drive that delivers you from A to B, but admit it, like me you have always wanted that classic ride that reflects your true personality. Well I am here to tell you, it's a real possibility and one I have achieved and would like to share with you.

My first classic car was an Australian designed and built 1974 Ford LTD ("LTD by Ford") sedan. The advertising touted it "Ford LTD… the classic limousine". How I loved that slogan! For Australia, they were marketed like a Lincoln would be in America, or a Jaguar in the UK. They were the pinnacle of locally built luxury in early 1970s Australia, where most cars didn't even feature a radio as standard The LTD - or Landau, the two-door version - was offered standard with a 5.8 liter V8 engine, power disc brakes, power windows, air-conditioning and even concealed headlamps. The only options were leather upholstery and stereo cassette player.

I was 34 years old, a year into a new profession and decided it was time to begin my search. A handful of monthly magazines listed For Sale ads with a black and white photo, or if the advertiser paid a premium, a color shot, together with a brief description of the car and a phone number. If you wanted to know more, you picked up the phone and spoke to the seller. One day, I spotted an Apollo Blue LTD, second owner, low miles and all original. That was the start of my journey, which continues today, 26 years and several cars later, but still living the dream.

My Custodianship:

- 1966 Ford Mustang 4.7L Convertible (2018 – Current)

INTRODUCTION

- 1974 "P5" LTD by Ford 5.8L Sedan (1998 – 2012)
- 1976 Rolls-Royce Silver Shadow 6.75L Sedan (2023 – Current)
- 1977 Ford Escort Ghia 1.6L Sedan (1982 – 1984)
- 1979 Chrysler Sigma/Galant SE 2.0L Sedan (2010 – Current)
- 1980 Chrysler Sigma/Galant SE 2.0L Sedan (1984 – 1994)
- 1990 Ford Mustang LX 2.3L Convertible (2021 – Current)
- 1994 Holden Commodore Exec. 3.8L Sedan (1994 – 2008)
- 2003 Ford Fairmont 4.1L Sedan (2012 – 2017)
- 2007 Holden Epica/Malibu CDX 2.5L Sedan (2008 – 2012)
- 2008 Kia Cerato/Spectra EX 2.0L Sedan (2008 – Current)
- 2008 Jeep Commander CTD Limited 3.0L (2017 – Current)
- 2009 Mercedes CLK AMG 3.0L Convertible (2022 – Current)

So why should you do the same? Because life is short, and if you are a classic car lover like me, you will reap tremendous rewards along the way and own a slice of motoring history that is admired the world over. Don't let your hesitations hold you back, explore your options carefully, follow your heart and your head and enjoy the ride. This book is designed to help you with the why, what and how to go about it.

2

Why Buy a Classic Car

Many people in your life will tell you classic cars are expensive, time consuming, difficult to maintain and even a pain the neck. Maybe they speak from bad experience, or maybe they simply don't share your passion. Either way, the decision is yours, and with some insights, tips and tricks, it needn't be any of those negative experiences. The goal here is to buy the best car you can afford, and one that will be a pleasure to drive, own and show decades into the future.

Classic car ownership is a passion shared by millions of people all over the world, so you will be far from alone on your journey. One of the greatest benefits comes from sharing your interest with like-minded enthusiasts who will inspire you. In fact, regardless of your ride of choice, "car guys" and "car gals" come to recognize each other as family, with whom you share a special and unique bond. Speaking for myself, very few of my close friends or family share my passion for cars, for most it's a means of transport – I call them "A to B ers". They just don't get it. But my car lover family definitely get me!

By now you have probably have an idea of what car you would like. You have even have researched clubs, Facebook groups, or other owners whom you can make contact with. If not, I would suggest you seek these people out. Begin to understand how you might become part of that community, what events may be held in your local area that you can attend, and how you can dip your toes in the water without spending a single cent. Get a feel for the lifestyle that comes with classic car ownership and the excitement of group interactions and showing off your pride and joy.

With classic car ownership also comes an intense feeling of satisfaction, pride and achievement. Belonging to a club for example, can inspire you to improve your car, attend competitions and even win trophies that you will cherish forever. The benefits and goals you set for yourself are quite literally limitless with this passion, it just depends how serious or competitive you want to be, and what budget you want to maintain. The point is, classic ownership can be anything and everything you want it to be. Keep it simple, or make it your ultimate investment in time and money, it doesn't matter. All that does matter is its fun, it's healthy and it's all in the spirit of keeping these beauties alive and on our roads.

Are you one of the many classic car lovers like me who look at new cars and think to yourself, don't they all look the same? Well they do! Remember when cars were designed by artists? When models were crafted in clay, when engines where built with power rather than economy in mind. Now I'm not suggesting we ignore the principles of a green planet, but we are not talking about your daily drive here, we are talking about an interest vehicle that fulfills your dreams and drives your passions. Does it matter that a car you drive on weekends is not the most fuel efficient? Not in the slightest! Think of it this way, we are recycling aren't we? The car already exits! There are no carbon emissions in the production of the car – it's most likely 50, 60 or even 70+ years old, that's done and dusted and in the past.

As for simple engineering, it is a huge plus for the novice mechanic. You aren't having to deal with elements of modern technology or computers that throw errors, or require another computer to service them. Old school engines generally work on a rudimentary carburetor, points and spark plug system that most home mechanics can tweak and most importantly, can be repaired with a minimum of fuss. No, I am not a mechanic myself, but rudimentary engine skills are all that is generally required for most popular classic car makes. My suggestion, is to familiarize yourself with basic engine and battery maintenance, find a reliable mechanic if you are not keen on doing the work yourself – there are plenty of resources out there – but most of all, don't let a vintage engine deter you. It is far easier to keep a simple combustion engine on the road than a high-tech one when it fails. Age and simplicity are your friends in this regard.

Another important aspect is investment potential. The cost of acquiring and maintaining your dream car is a major factor, no doubt about it, but try to remember that it's effectively an investment. Your friends and family may argue this point, but at the end of the day, you are not spending your hard earned cash on a throw-away item or an asset that easily depreciates. Classic cars rarely go down in value, rather they tend to increase, and especially so when you invest further in them. Naturally it depends on exactly what you buy, but if you buy sensibly, the market value for your classic vehicle will only increase if you look after it diligently and store it wisely.

3

What to Buy

My guess is you already have your heart set on the car you always wanted. Am I right? If so, I'm not about to change your mind, and nor should I. Your dream is your dream and there are few substitutes. My dream car is a 1969 Mercury Marquis Convertible. Sadly no, I don't have one in my garage. They are very rare – at least in mint condition – and I have never come across one that I felt committed to. To this day, I lust after one, but it will have to be the one that got away. Remember, I live in Australia and I need to consider all the factors that you too will be considering, to make it a viable option. Which is what I'm about to cover in this chapter. What to buy is governed by many factors, but I reiterate, keep your dream in mind!

One of the key factors will be budget. You will know what you're aiming for versus what you have in the bank as disposable income. However, don't let limited savings get in the way either. Yes, we need to budget – even for our dreams – but you may have the option of a car loan to bolster your spend. Keep in mind, this will be an investment too, as we talked about earlier. In fact, if you think about it, a car loan for a classic could be more viable that a car loan for a new model. What do new cars do as soon as you drive them from the dealer's lot? They depreciate… a LOT! This is something classic automobiles rarely do. They maintain their purchase price if cared for, and if anything, they increase. Wouldn't you rather take a loan for an appreciating asset – and one that you can further value add to – rather than a depreciating asset? I know I would. So a moderate loan is not out of the question. When talking to your broker, explore the options for a specialized classic car loan, as some lenders take the appreciating factor into account.

No doubt you already have an awareness how asking prices range for your new pride and joy. The thing to remember here is "you get what you pay for". This is never truer than for a car… any car! I always remember The Brady Bunch episode when Greg buys the '56 Chevy Bel Air convertible (think how much that's gone up in value!). Mike tells Greg "Caveat Emptor", or "Buyer Beware". If you are like me, you'd rather spend more up-front and save more down the track, which is what Greg didn't do. He was talked into a cheap car that looked okay, but under the covers, it was cheap for a reason. Very rarely do you find a bargain deal that is a true bargain and a gem at that. It happens for sure, but those occasions are few and far between, no matter what the purchase. My suggestion is to take your maximum budget and work from that figure. I would also suggest that whatever your maximum figure is, you add another 25% to the price of the car to allow for post-purchase expenses. I have never bought a car that I am 100% happy with; I always have to do something to make it my own. "turn-key" is a great marketing phrase, but rarely a reality.

Talking about additional expenses to factor in, you will need to consider where the car is located. Is it within driving distance of your location? If not, do you need to transport it, or do you need an airfare to fly there to collect it and then drive it home? There are also registration fees in your State, certificate of road worthiness, government stamp duties, car insurance etc. Make a list of what expenses you will encounter for the car you are considering and add all these costs into the maximum figure you plan to spend. Are you still within budget? Keep in mind that these ancillaries are not negotiable, only the price of the car may be open for discussion.

You will also need to consider availability and whether there are examples for sale in a condition that you consider acceptable. Availability will dictate price to some degree. If there are several options you would be happy to buy, it may be a buyer's market for your car, but if not it probably falls within the "rare" category and this will drive up the price tag. An example of a car in abundance that still commands a premium is the original Ford Mustang (1964½ to 1966). Twenty years ago, these could be had for couch change, but now there is a global awareness of their popularity – almost a cult following – so this is of course reflected in their current value. Whatever you have your eye on, look at the quality of what you're getting for your money and what you need to spend to put your stamp on it, plus the non-negotiable costs to get it home and driveable.

On-going availability of parts is possibly the greatest consideration in the entire decision making process, and that is why I sold my cherished LTD in 2012, after fourteen years proud ownership and several trophies. Being a design unique to Australia and built in very limited numbers for the few who could justify their expense, Australian Fairlane and LTD are difficult to maintain in mint condition. A single accident – even minor – can render these beautiful cars almost impossible to repair. I never wanted to face that situation head on, so I decided to let her go to her next keeper.

This is my humble advice to anyone considering a rare species – investigate parts stock before you buy. Connect with Facebook groups, other owners, join a marque club for advice, and scour the Internet for used and after-market suppliers, so you are fully aware of any challenges in advance. This includes mechanical parts, body panels, lenses, interior and exterior trim and moldings, badges, plus any consumables. Naturally where there is a will, there's a way, but I can say from experience, it can take days, weeks or even months to track down a part that the seller knows all too well is a rarity. Alternatively, with 3D Printing, a plastic part can be re-created, but not if the original is missing, or significantly damaged. For American and Canadian classics, you may take comfort in the knowledge that many cars within the GM, Ford or Chrysler family shared numerous parts with their siblings; therefore your scope of availability is considerably wider than for the rest of the automotive world.

Classic cars are offered for sale in all manner of states, from "project car" examples to fully "turn-key" (there's that term again!) restorations. You will be asking yourself, how much time, effort and cash am I prepared to invest in this piece, and if she needs work, can I do this work myself? A project car will demand your time like a mistress, she will also demand and arsenal of tools, a ramp or perhaps a hoist, adequate lighting, power and water. Even the turn-key example will require all of these assets to maintain her in the condition you have purchased her in, only the time investment will be less. Maintainability is the key to preserving your collector car for future rewards. Nothing would be worse than spending your hard-earned savings, only to see her fall into a condition of neglect.

Using my Rolls-Royce Silver Shadow as an example, this is not a car I am experienced enough to maintain in every aspect of her health. Perhaps your classic purchase will likewise require a specialist mechanic or technician. That's okay too, so long as you have a specialist in mind for more complex issues, and they are close enough to make house-calls, or you can drive or trailer the car to their shop. Alternatively, if you a brave enough to tackle even the trickiest of problems, be sure to have a support group of other owners you can call or message for advice and guidance. This is where Facebook or other club oriented groups will be invaluable.Most fellow owners will be more than happy to share their knowledge with you, remember they share your passion for the marque!

You also need to ask yourself, where she will be stored for either restoration, or preservation. I have seen many would-be buyers wide eyed and excited about their new ride, only to realize they can't accommodate it appropriately. Investigate whether your garage is large enough to house her along with all the gadgets and devices necessary for her well-being. If you don't have this storage at your home, is renting an off-site storage shed an option? It certainly is my solution for several of my cars – I have four stored at home (including my daily drive), and three stored off-site at a secure facility, with access to power, water and additional work and storage space. If you need off-site storage, add the monthly fees into your overall budget.

Lastly we cover the angle of potential resale value and demand. You were attracted to a particular make and model, so there's likely a keen demand out there by other enthusiasts. Providing you have not over-spent on her acquisition, or over-capitalized on her restoration or customization, chances are you will be able to re-sell her too, when the time comes. The question should be asked though, could you break even with your outlay, might you profit from your investment, or will you likely lose money if you sold her either unfinished, or in a state of neglect? Consider these points as they might be relevant to your long-term plans for the car.

I have never bought a car for the purpose of flipping for a profit. That's probably a topic for another book. Each car I have bought has been undertaken with a "collectors" view, to satisfy my desire to keep a car I have always lusted after – sadly for my bank manager, there are several! My investment in each and every one of them has been for my own personal interest, satisfaction and pride of ownership. That said however, I have stored them correctly, maintained them to a high standard and given their rarity as time marches on, they will serve me well enough when the time arrives to part ways. Of that I am confident.

4

How to Buy

Times and technology have changed overwhelmingly since my first classic car purchase way back in 1998. Today we enjoy a plethora of options to source not only the cars themselves, but the parts we need to maintain them. You have at your fingertips more resources than your fathers or grandfathers ever had in their day. Use these resources to your best advantage. Research is the key to locating that (near) perfect example you have dreamed of. Here we can explore those options in more detail.

My first recommendation is to join a local or regional club that represents the marque of your choice. For example, a year prior to buying my Rolls-Royce, I joined the RROCA - Rolls-Royce Owners Club of Australia, as an associate member (member without a car). This opened up doors to vehicles for sale, mechanical services, parts sources and even options for reduced insurance and registration fees upon purchase. Clubs and groups exist all over the world to help you with your purchase and to support you on your classic car journey. They always welcome new members who share the dream. The low membership fees for such clubs will be worth the investment in the long run.

Internet auction sites are often our go-to resource for all manner of goods nowadays, and classic car auctions are no exception. You have access to stock from literally all over the world. Represented here are both private sellers and dealer services; photos and descriptions tend to be excellent and you can compare and contrast fixed prices and auction activity. The only thing you can't do, if the car is not local to you, is physically inspect it yourself. Now this doesn't matter so much for a moderate purchase, particularly of a new item, but for a classic vehicle, it can be a mine-field. Don't get me wrong, I have purchased several of my cars using exactly this method, but in each case I have made contact with the seller and exercised due diligence in the process. If you can view the car yourself then great, if not, arrange for a friend or inspection service to do this for you. This is a bare minimum! Try not to get carried away in the excitement and hit "Bid" or "Buy Now" before knowing exactly what awaits you. Very rarely can you return a car for a refund once the deal is done and buyer protections often exclude such purchases. Do your research!

The live auction is another ball game altogether. As distinct from a "timed" internet auction, the live auction comes with time and commitment pressures, plus a hype of excitement designed to garner buyer competition and ultimately drive up the bidding. If you know the lot is listed well in advance, you can perform your due diligence similar to a timed auction, but often you will not have access to the vehicle for pre-inspection. If you cannot attend the live event yourself, you may opt to employ one of the auction house's proxy agents to bid for you, within your budget limits of course. But beware of getting caught up with the excitement and stay within your upper limit. Don't be concerned about dropping out if your target lot exceeds your price expectations, there is always another car. In summary, this is not my preferred method of purchase and I have never ventured down this path. Live auctions are great to watch for entertainment and more often than not feature some extraordinary vehicles, but they are simply not for me as a cautious buyer.

Specialist classic car dealers are another option to find your ride, again represented on the internet and more than likely participate in the auction space as well. These sellers offer a little more protection than a private seller, but only marginally in my opinion. The only advantage with a dealer is they have – or should have – a greater responsibility to deliver what you pay for, plus you have access to reviews about their business practice. Always read these reviews, take half of what you read and believe only half of what remains. They are a guide only; each and every buyer will have their own experience with a given seller, and remember every classic car is totally unique. Be sure to talk with the dealer, either in person, or over the phone. Establish a rapport with one person and have a list of questions that are important to you. It also goes without saying that a sticker price is just that – a "serving suggestion" if you like. Don't be afraid to make an offer lower than the asking price, even if the dealer suggests it's already a bargain. My personal experiences with dealer sales have been positive, especially so for classic car dealerships. If they trade in this specialist field, they are more than likely a car enthusiast also.

Some of the best buys can be had through a private seller, including that elusive unicorn – the Barn Find. This is where you can potentially snag a one-owner and/or low mileage prize with a full set of documents and manuals, original build sheet and keys, and sometimes original tires and wheel covers. Often the sale is the result of the owner's retirement or passing, and very often the car presents in factory "stock" condition, having been cherished and garaged its entire life. Now that is a find! Keep in mind when viewing dealer and auction examples, these cars sometimes find their way to the dealer or auction block by way of a private seller or deceased estate seeking a fast and efficient disposal. However, if you are not purchasing directly from the seller without a middle man, the information you receive may be inaccurate or incomplete. As with dealer sales, a purchase from a private party is quite often negotiable, unless the seller has specified "firm" in their asking price.

If your plan is to customize your ride, a stock, original car may not be your goal. Indeed it probably is not, as the premium for these original unicorns may not be worth spending, particularly if you intend to modify it with up-rated engine, custom paint, wheels, exhaust, lighting and the like. I'd go as far as to say leave the unicorns for those who are prepared to indulge their ultimate nostalgia, so they can continue to preserve and present the car as it was when it rolled out of the factory back in 19xx. In essence, these cars are, or can become, museum pieces in their own right.

Now let's talk the 'R' word – Rust! This is by far the worst demon to avoid at any cost, if you possibly can. Not a single car is immune to metal cancer, no matter how well it's been bubble wrapped, particularly in areas where snow is prevalent, or salty sea-side locations. And let's not delude ourselves; even brand new cars have commenced their degradation right from the factory floor. The key is to minimize its impact on your wallet. We all know the magnet trick to test the exterior panels and our eyes should be trained on unusual bumps or bubbles in paint work, but rust finds its footing on surfaces in far less obvious places too. You can't pull the car apart in search of the dreaded brown crust, but do yourself the very best favor and seek professional advice if you are in any doubt what so ever.

To share an experience I don't want to re-live, my 1966 Mustang concealed a nasty secret. This particular car was located 1,100 miles/1,700 kms away and I was unable to inspect her prior to purchase. Exercising my due diligence, I arranged for a classic car specialist to inspect her for body, paint, mechanical and electrical issues before the deal was closed. The inspector's report was for most part favorable, with the exception of some minor flags that the seller had already disclosed. Good to go right? Well, I thought so, until I took delivery of her, and months later started fishing in places that were clearly overlooked. Unfortunately, the convertible top roof well behind the rear seat was heavily decayed and required considerable fabrication and repair works to reinstate. Needless to say, this was a costly exercise and a hard lesson learned. Fast forward several years and tens of thousands of dollars, she is as good as new – in fact museum quality – but I'd have walked away if I'd known what was lurking under the covers before buying. Knowledge is power and it can save you life-changing amounts of dollars.

Lastly, a caution about insurance, as this forms part of the buying and budgeting process. In many parts of the world, including Australia, specialist insurance companies exist to cover your classic. Their policies are designed to cater for the unique needs of car enthusiasts, with special coverage for lay-up periods, including restoration and museum loan-outs, agreed payout value, choice of repairer, club use only discounts etc. You are advised to seek out these suppliers and consider all their options, making sure whatever you purchase is fully insured for purpose, prior to taking delivery. I cannot stress enough the importance of this step. So many owners overlook this necessity or postpone it, but you are leaving yourself wide open to disaster from a personal and even legal stand-point.

5

The Joys of Ownership

Now for the best part! You have worked hard to arrive at this destination. All the investigation, analysis, ticks and balances have been performed. Let's talk about the ultimate experience, what pleasures await you and the dream car in your garage.

My very favorite pass time is a simple one. It involves just me, my car, a tank of fuel, the soundtrack of my engine and a sunny Sunday morning. For you, it may include your partner, your family, friends or even a group of like-minded enthusiasts on a club drive. I doubt you will drive your classic every day, so this time is precious and will be filled with pride and elation. This is a significant achievement for you, the culmination of potentially years or decades of research, to relish in whatever form it takes. You will receive glances of admiration from other motorists, plenty of thumbs-up, impromptu chats at service stations and car parks, and ultimately be the envy of so many other folk who wish they were in your driver's seat.

At some point you will likely be asked to don your chauffeur's cap and donate your services for a family or friend's wedding, your kids' high-school prom or even a charity event, when you have the opportunity to put smiles on envious faces. These occasions are just cream on the cake. You play an important role within your social circle and in your local community with these gestures, that mean so much, yet cost so little. Let's face it, you'll want to grab every opportunity to drive your pride and joy, so why would you decline. These celebrations will be etched in your memory for life, plus in the photos and movie collections of all those around you who participate.

Club runs and car shows feature also heavily on the menu for owners of all classic vehicles. They form an important part of your ownership experience, where you can engage with fellow owners, share top tips, source parts, and network with other clubs and suppliers of products or services you may need in the future.Never was the term "it's not what you know, but who you know" more prevalent than in the world of classic cars. There is a wealth of information out there, much of which is best exchanged in person at events like these.Take part in as many as you can. They are fun, informative and rewarding in every respect. Keep an eye out for up-coming events in your area, notices are published well in advance and most are posted on-line.

Of course the "golden egg" and ultimate accolade any enthusiast can attain is a trophy. Enter your car into the right show, aim for your category and you too will drive home with an award equal to your effort. Different categories include best original, best restored, best overall, best in year (range), or the highest of them all, the concourse winner, plus many other awards depending on size and nature of the show. An unspoken, less tangible advantage to any trophy or ribbon is the cache' of owning a trophy winning vehicle and with that, the value of your classic invariably increases. When you seek out the For Sale listings, how many times did you notice "trophy winner" in the advertising? The reason is obvious – not only is the car of a certain standard, but it's been officially recognized by a judge or judges to that effect, and in turn, the inherent value of the vehicle rises.

Another source of unashamed kudos is the Museum loan-out, or fixed-term donation. Motor exhibitions and their curators rely on displays that change periodically to maintain visitor interest year-round, therefore your art on wheels can be an attractive proposition to them. Talk to your local or regional motor museum, gauge their interest and register your ride for an upcoming display. This is an ideal way to show your stuff in a safe and controlled manner. Your car may feature in advertising for the event, it is securely housed and maintained, and upon return, you are in possession of a bona fide museum piece. Once again, your ride's market value may have increased immeasurably.

Finally, if you are inclined to offset your ownership expenses or even profit from your classic, you may consider registering with an advertising, television or movie props agency. Star cars are real business and a lucrative one at that. But be prepared for an absence of your vehicle, possibly shipment to another area where filming takes place and for others to drive and handle your car. Make sure your insurance or that of the agency or establishment using your asset is fully covered for any event. The possibilities in this space are endless, particularly if you reside in a location where higher budget contracts are on offer.

For the rest of us, we simply want to enjoy our classics when and however we can. There is reward enough just being their "keeper" for a good time, a short time or a life time. I use the word "keeper" or interchangeably "custodian" because most, if not all of these treasures, will outlive us and be handed down as heirlooms, on-sold to younger generations and continue to represent their place in time when life was simpler and design was considered first and foremost by their makers.

6

Conclusion

By now you will have a much clearer understanding of what steps are involved in your classic car journey. My message to everyone is to enjoy the experience from your initial research, right through to that first Sunday drive, but most of all, to undertake the process with as few pitfalls as possible. There will be speed bumps along the way, but we all encounter those in one form or another. Follow your heart AND your head, manage your costs responsibly, and don't let the naysayers talk you out of what I truly believe to be one of the greatest privileges life has to offer. I get it!

I sincerely hope this candid but informative guide has achieved its purpose by helping you, or a family member, to navigate classic the car ownership experience. I can promise you that with all angles and possibilities considered in this book, and with due diligence as your mantra, you will reap the unbounded rewards of custodianship long into the future.

Good luck and happy motoring!

www.ingramcontent.com/pod-product-compliance
Lightning Source LLC
Chambersburg PA
CBHW050254230526
45470CB00005B/2267